The Open University

Open Mathematics

Unit 4

Health

MU120 course units were produced by the following team:

Gaynor Arrowsmith (Course Manager)
Mike Crampin (Author)
Margaret Crowe (Course Manager)
Fergus Daly (Academic Editor)
Judith Daniels (Reader)
Chris Dillon (Author)
Judy Ekins (Chair and Author)
John Fauvel (Academic Editor)
Barrie Galpin (Author and Academic Editor)
Alan Graham (Author and Academic Editor)
Linda Hodgkinson (Author)
Gillian Iossif (Author)
Joyce Johnson (Reader)
Eric Love (Academic Editor)
Kevin McConway (Author)
David Pimm (Author and Academic Editor)
Karen Rex (Author)

Other contributions to the text were made by a number of Open University staff and students and others acting as consultants, developmental testers, critical readers and writers of draft material. The course team are extremely grateful for their time and effort.

The course units were put into production by the following:

Course Materials Production Unit (Faculty of Mathematics and Computing)

Martin Brazier (Graphic Designer)
Hannah Brunt (Graphic Designer)
Alison Cadle (TEXOpS Manager)
Jenny Chalmers (Publishing Editor)
Sue Dobson (Graphic Artist)
Roger Lowry (Publishing Editor)
Diane Mole (Graphic Designer)
Kate Richenburg (Publishing Editor)
John A. Taylor (Graphic Artist)
Howie Twiner (Graphic Artist)
Nazlin Vohra (Graphic Designer)
Steve Rycroft (Publishing Editor)

This publication forms part of an Open University course. Details of this and other Open University courses can be obtained from the Student Registration and Enquiry Service, The Open University, PO Box 197, Milton Keynes MK7 6BJ, United Kingdom: tel. +44 (0)845 300 6090, email general-enquiries@open.ac.uk

Alternatively, you may visit the Open University website at http://www.open.ac.uk where you can learn more about the wide range of courses and packs offered at all levels by The Open University.

To purchase a selection of Open University course materials visit http://www.ouw.co.uk, or contact Open University Worldwide, Walton Hall, Milton Keynes MK7 6AA, United Kingdom, for a brochure: tel. +44 (0)1908 858793, fax +44 (0)1908 858787, email ouw-customer-services@open.ac.uk

The Open University, Walton Hall, Milton Keynes, MK7 6AA.

First published 1996. Second edition 2008.

Copyright © 1996, 2008 The Open University

All rights reserved. No part of this publication may be reproduced, stored in a retrieval system, transmitted or utilised in any form or by any means, electronic, mechanical, photocopying, recording or otherwise, without written permission from the publisher or a licence from the Copyright Licensing Agency Ltd. Details of such licences (for reprographic reproduction) may be obtained from the Copyright Licensing Agency Ltd, Saffron House, 6–10 Kirby Street, London EC1N 8TS; website http://www.cla.co.uk.

Edited, designed and typeset by The Open University, using the Open University TEX System.

Printed and bound in the United Kingdom by The Charlesworth Group, Wakefield.

ISBN 978 0 7492 5432 2

Contents

Study guide	4
Introduction	6
1 Patterns	7
1.1 Cholera in Soho	7
1.2 The passionate statistician	9
1.3 Smoking and lung cancer	10
1.4 Leukaemia clusters	13
2 Chance	17
2.1 Random numbers	18
2.2 Investigating variation in random numbers	20
2.3 Hot spots	23
3 A pattern isn't proof	26
4 Two tales of variation	29
4.1 A mother's tale	29
4.2 A farmer's tale	34
Unit summary and outcomes	38
Comments on Activities	40
Index	45

Study guide

This unit consists of four sections, which are best studied in order but, if necessary, the video can be watched later. They are not all of equal length. Section 1 aims to start you thinking about patterns in data and what conclusions can be drawn from them. Section 2 is longer than average and includes a subsection based on a video band called 'Hot spots'. Section 3 is very short; it summarizes briefly the main ideas discussed in the first two sections. Sections 1, 2 and 4 each involve working through a section from the *Calculator Book*.

As you work through the unit, you will be developing and applying mathematical skills such as drawing and interpreting scatterplots, calculating measures of spread and making a tally chart. You will have the opportunity to generate random numbers using your calculator and to investigate any patterns that result.

You will be asked to consider your response to two newspaper articles which draw rather different conclusions about the same issue.

As well as considering the newspaper articles and providing comments on them, take the opportunity to consider *how* you are reading the unit for learning. Many people take the skill of reading for granted. But you should not forget that technical reading is a skilled activity, requiring making sense of often complex material.

As an Open University student, you need to do more than simply read, because you may be asked to extract information from a range of different types of material—for example, from tables, figures, a calculator book, even a video!

There are two television programmes associated with the themes of this unit. The first, *The Passionate Statistician*, looks at the life and work of Florence Nightingale and is relevant to Section 1. The second, *Asthma and the Bean*, offers a case study detective story explaining possible causes of an outbreak of asthma in Barcelona and follows on from the discussion of Section 3. It is also relevant to *Unit 5*.

As you work through Block A, you need to think about how and when you will complete the assessment—both the CMA and TMA questions. In your study plan, try to include time that you can devote to looking at and completing the relevant assignment questions.

At this stage, you may already have received some marked work from your tutor. Assignments are not designed only for assessment—although most count towards your final score. Working on an assignment is part of your learning of mathematics, but to get the full benefit from a TMA, you will need to consider your tutor's comments and think how you can use them to make improvements. Are there particular areas where you want to make changes, or skills you want to concentrate on in completing your next written assignment? Are there any concepts you need to review?

Included in your Learning File are two sheets relevant to planning and evaluating your work. At some stage during your study of this unit, take some time to think about how you intend to study the rest of this block and complete the relevant parts of the assignment. First, use the 'Evaluation' sheet to review your work so far, including any comments you have received from your tutor. Think about one or two areas that you want to pay attention to, for instance, organizing your time better, or not leaving the assignment until the last possible moment. Now use the 'Planning' sheet to note down how you intend to complete the assessment (both the CMA and TMA) and your study of *Units 4* and 5. Include any areas you aim to concentrate on that you noted on the evaluation sheet. Keeping track of your work, prioritizing, monitoring and reviewing progress are important skills in becoming an independent learner. By completing the planning and evaluation sheets you are keeping a record of your progress.

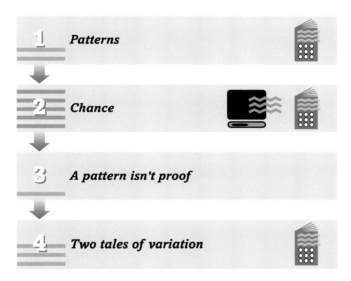

Summary of sections and other course components needed for *Unit 4*

Introduction

It is sometimes said of a person who is suffering from the common cold that, with proper medication, it can be cleared up in a week but, if left untreated, it can drag on for seven days!

Many who attend a doctor's surgery like to depart clutching a prescription of some sort—the prescription is at least a sign that someone of importance has acknowledged our ailment and has taken steps to remedy it. But how do you know whether the pills or medicine you swallow or the ointment you rub on really do any good? It is possible that, left untreated, the ailment would have cured itself anyway. This is the main reason that 'quack' cures have flourished in every society over the centuries—cause and effect, particularly in the area of health, is extremely hard to prove or disprove. You might say with conviction that 'it seemed to work for me', but when it comes to advising others about cures, most of us can only draw on a rather small sample of personal experiences.

> The word 'quack' is an abbreviation for 'quacksalver', meaning an individual pretending to possess medical skills.

This unit is about clusters of disease and how they can be interpreted. The central question of interest here is whether an apparent cluster is sufficiently marked to warrant further investigation. But sometimes groupings appear just by chance. You need an understanding of just how likely these apparent groupings are if you are to make sensible judgements about 'patterns' in observed data. So a major element in your study of this unit will be to explore patterns which arise by chance. You will do this first with your calculator and then you will see a larger-scale, computer-based investigation in the video section.

In Section 4, you will be able to draw together and extend ideas about variation and measures of spread in the context of the birth weights.

Activity 1 Reading for learning mathematics

As you work through this unit, you are asked to think about *how* you are studying and learning mathematics using the different course components. This is not an easy or straightforward task, so do persevere. Some questions that you might consider as you 'read' the unit are:

◇ Are you clear about what your purpose is in reading at any given point?

◇ If you are unsure of something, what do you do to clarify the meaning?

◇ Do you find it easy or difficult to extract the main points?

◇ What techniques do you use to help identify these main points?

Record your initial responses and any other comments on the printed response sheet.

At the end of each section of the unit, add further comments to this sheet.

1 Patterns

Aims The main aim of this section is to illustrate how a perceived pattern in data can suggest possible cause and effect relationships. ◇

This section consists of brief descriptions of patterns in data which have arisen in investigations of illness and disease. One key idea to focus on in each account is the particular cause and effect relationship that may be at work and how the statistical information is related to this.

1.1 Cholera in Soho

In the middle of the nineteenth century, before the theory of infection by germs was established, the means by which cholera was transmitted was not understood. John Snow (1813–58) was a distinguished doctor of the time: he was Queen Victoria's obstetrician and a pioneer of the scientific study of anaesthesia. He investigated the role of drinking water in the spread of cholera and became convinced that contaminated water was responsible. In 1849, following the London cholera epidemic of 1848, he put forward the theory that cholera was caused by 'a poison extracted from a diseased body, and passed on through drinking water which was polluted by sewage'.

In 1854, there was another major cholera epidemic in London. Snow noted the location of all the cholera cases in Soho and plotted them on a map. A portion of this map is reproduced overleaf. The bars represent fatal cases in each household.

Activity 2 Polluted pump

The map shows the location of water-pumps in the vicinity. Do you think any or all of the pumps shown on the map could be implicated in the outbreak of cholera?

Comments on Activities start on page 40.

Both the epidemic of 1854 and the earlier epidemic of 1848 were particularly severe in Broad Street. In 1854, for instance, there were deaths in all but twelve of the forty-nine households in the street. Snow found that all the cholera victims had drunk water from the Broad Street pump, whereas those who escaped the disease had obtained their water elsewhere. Using the map to back up his assertion that the pump was contaminated, he was able to persuade the authorities to put the pump out of action. The residents were obliged to seek other water sources, and this helped to control the epidemic.

UNIT 4 HEALTH

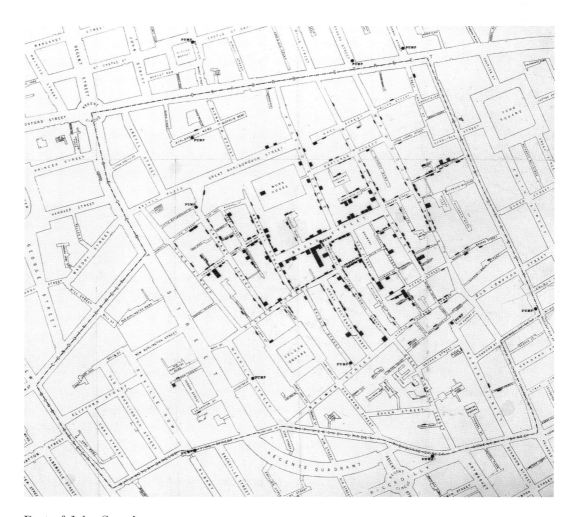

Part of John Snow's map

By looking at the *pattern* of where the victims lived and where they obtained their water, Snow was able to demonstrate the likely source of the infection. This, in turn, led to a better understanding of the disease transmission mechanisms. Maps are a useful way of presenting a considerable amount of information so that patterns can be identified.

Activity 3 *Alternative explanations*

The fact that there were more deaths near the Broad Street pump than near the other pumps in the area suggested to Snow that the pump might be implicated in the epidemic, but did not *prove* that this was the case: the actual mechanism for the spread of the disease was not known at the time. Can you think of another explanation for the greater numbers of victims to be found in the neighbourhood of this pump?

Given only the map and no other information, another explanation might be that there was a source of infection in the locality of the pump, but not associated with it. Another point to bear in mind is that no indication of the population density is given on the map. This raises two separate but related issues. First, the greater number of cases could be due to there being more people living in Broad Street than in the neighbouring streets. And second, if the population density were high in this locality, it could be that there were many families living in high density accommodation in relatively poor conditions: in such conditions, the disease might spread rapidly. The point to note here is that the cluster evident from the map did not *prove* that the Broad Street pump was the source of the infection in the locality (although this was eventually found to be the case). However, it did suggest what turned out to be a fruitful line of investigation. As a result, a much better understanding of the disease was gained, namely that it is caused by water-borne bacteria. This eventually led to the virtual eradication of the disease in many parts of the world. Today, cholera outbreaks are mostly confined to areas where there is poor sanitation and an inadequate supply of clean drinking water.

The next example is also historical.

1.2 *The passionate statistician*

Florence Nightingale (1820–1910) is perhaps best known as 'the lady with the lamp' who tended the injured soldiers in Scutari hospital during the Crimean War (1853–1856). Coming from a wealthy family, Nightingale was potentially marked out for the relative obscurity of Victorian marriage and maternity. However, she believed that she was called by God for greater things. In her early thirties, she announced herself to be 'up to her neck in china and linen' and shocked her family and friends by setting off to tend 'the great unwashed' in the Crimea. At that time, so-called 'common' soldiers were treated appallingly by their officers and to be nursed lovingly by a great lady was a moving experience for them. It was said that 'even the rough soldiers kissed her shadow' as she passed by on her rounds.

After returning from the Crimea, Nightingale became a public heroine and a personal favourite of Queen Victoria. Shunning public acclaim, she retired to her bedroom and from there turned her energies to the reforms of hospital hygiene and their design. What is less well known about Florence Nightingale, however, was her lifelong interest in, and knowledge of, the subject of statistics.

Most of the accounts of her life and work are agreed that she cared passionately about the plight of others—particularly those unable to fend for themselves, like the poor and the sick. She took the view that the ruling classes of Victorian England (both in the military hierarchy and in government) were either unwilling or unable to see the truth about conditions in Scutari and elsewhere unless presented with irrefutable evidence. Her life's work as a statistician, therefore, centred around two key themes. First, she needed data that she could trust. The data had to be accurate, comprehensive and comparable between hospitals and between regions. This resulted in her long-running campaign to standardize hospital record keeping.

Second, she was aware that most of the decision makers around her were statistically illiterate and tables of figures, however accurate, were simply not winning over their hearts and minds. This led her to devise creative and innovative forms of diagrammatic and graphical representations of her data which demonstrated the underlying patterns in a way that no one could ignore.

> This would be a good time to read the reader article 'Florence Nightingale' by I. Bernard Cohen.

You will find examples of her diagrams in the reader article about her life and work. The title of the television programme, *The Passionate Statistician*, captures an apparent paradox: how someone who cared passionately about people could devote her life to statistics, which many people see as being dry, dusty and lacking in passion.

The paradox is resolved by an understanding of Florence Nightingale's strategy for achieving social change. She believed that logical, factual arguments would win the day, *provided* they were based on valid data and were presented in a way that the decision makers could understand.

This brief description of Florence Nightingale's work is included here to illustrate how evident patterns in data have been used to provide insights into health and to influence policy making.

1.3 Smoking and lung cancer

The next example to illustrate relationships in health concerns some data which were published before the link between cigarette smoking and lung cancer had been established.

Today it is accepted by most people that smoking increases an individual's chance of developing lung cancer. But this link has not always been recognized and is still not accepted by some. A hundred years ago, very few people regarded smoking as a major health hazard; the possible link between cigarette smoking and lung cancer was hardly discussed at all before the Second World War. Carefully planned studies into the effects of smoking on health were only proposed after that war. The results of these studies have led to the situation today where (in the UK) smoking is officially accepted as being harmful to an individual's health.

Much was done to establish the link between lung cancer and cigarette smoking during the 1950s. The data in Table 1 were published in 1955 by Sir Richard Doll. The table gives the per capita consumption of cigarettes in 1930 for eleven countries and the male lung cancer death rates in 1950 in those countries. The per capita consumption of cigarettes is the average (that is, mean) number of cigarettes smoked per person in 1930. The male lung cancer death rate is the number of deaths from lung cancer per hundred thousand men in 1950.

Table 1 Per capita cigarette consumption (1930) and male lung cancer death rates (1950)

Country	Per capita cigarette consumption (number per year)	Male lung cancer death rate (per 100 000 of population)
Iceland	220	58
Norway	250	90
Sweden	310	115
Canada	510	150
Denmark	380	165
Australia	455	170
United States	1280	190
Holland	460	245
Switzerland	530	250
Finland	1115	350
Great Britain	1145	465

The data in this table are *paired data*: there are *two* items corresponding to each country—the mean cigarette consumption in 1930 and the male lung cancer death rate in 1950.

▶ Is there a relationship between the mean cigarette consumption and the male lung cancer death rate twenty years later?

It is difficult to tell from the table. But, as you have seen in *Units 2* and *3*, it is often easier to spot patterns in data from a diagram than from a table.

A *scatterplot*, sometimes called a *scatter diagram*, is a way of representing paired data. The data for Iceland and Finland have been plotted in Figure 1. The point for Iceland, for example, is the point which is above 220 on the horizontal axis (labelled 'Mean cigarette consumption/number per year') and level with 58 on the vertical axis.

In the next activity, you are asked to complete the scatterplot. Then use your calculator to obtain a scatterplot for the data, and check that you have plotted the points in the correct positions.

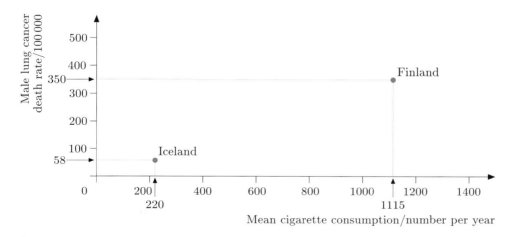

Figure 1 Part of a scatterplot of mean cigarette consumption and male lung cancer death rate

Activity 4 Completing the plot

Complete the scatterplot in Figure 1 by adding points representing the data for the other nine countries. Plot the points as accurately as you can and label each point with the name of the country to which it refers.

Now work through Section 4.1 of Chapter 4 in the Calculator Book.

Activity 5 Looking for patterns

Figure 2 shows the completed scatterplot for the data in Table 1. Describe any useful conclusions that you feel able to draw from the plot.

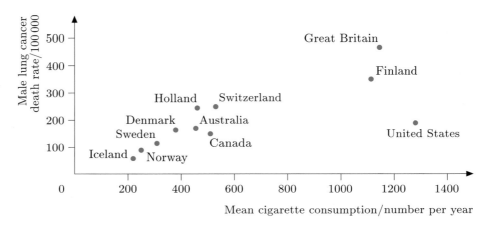

Figure 2 Scatterplot of mean cigarette consumption (1930) and male lung cancer death rate (1950)

Looking at the scatterplot, you can perhaps see that, generally speaking, the countries with low mean cigarette consumption in 1930 had lower male lung cancer death rates in 1950 than those where cigarette consumption was higher in 1930. There is a tendency for male lung cancer death rates and mean cigarette consumption to increase together. For example, the country with the lowest mean cigarette consumption, Iceland, had the lowest male lung cancer death rate. Whereas Great Britain, which had the second highest mean cigarette consumption, had the highest male lung cancer death rate. For countries such as Denmark and Switzerland, where the per capita cigarette consumption was in between the values for Iceland and Great Britain, the male lung cancer death rates were also in between.

It is worth noting that, with the exception of the two North American countries (Canada and the USA), the higher the per capita cigarette consumption of a country, the higher the male lung cancer death rate. You can see this either from Table 1 or from the scatterplot in Figure 2. The fact that the data for Canada and the USA do not fit the apparent pattern in the rest of the data is itself interesting. If there *is* a link between the per capita cigarette consumption and lung cancer death rates, then why should the male lung cancer death rates be so much lower for these two countries than for other countries with similar per capita cigarette consumption?

There might be any number of explanations for this. For example, perhaps in North America it was not customary to inhale the smoke from cigarettes as deeply as elsewhere; or perhaps more of the cigarettes had filter tips (if that makes any difference); or perhaps the cigarettes smoked contained relatively low levels of tar or nicotine.

Another possibility becomes apparent if you notice that, while the data on cigarette consumption are given for the whole population, both men and women, the lung cancer death rates are for men only. If data were available for the per capita cigarette consumption of men alone, then any pattern in a scatterplot of per capita cigarette consumption of men plotted against male lung cancer death rates might be different from that in Figure 2: the positions of the points for Canada and the USA might not be unusual on such a scatterplot.

This might be the case in the following situation. Suppose that, in 1930, North American women smoked more than women in the other countries. If this were so, then a smaller proportion of the cigarettes smoked in North America would have been smoked by men. In other words, it is possible that the per capita cigarette consumption *for men* in the USA was lower than in some of the other countries such as Great Britain and Finland; and, similarly, the per capita cigarette consumption *for men* in Canada might have been lower than in countries such as Holland and Australia. Without further investigations, you cannot know the reason for the 'lower than expected' male lung cancer death rates in Canada and the USA.

This is another example of a situation where a pattern in the data suggests a *possible* cause for a disease. This does *not* prove that cigarette smoking causes lung cancer: there could be other reasons for the apparent relationship. For instance, both cigarette consumption and lung cancer might be linked to the level of stress in society. Or smokers might work in conditions which encourage lung cancer. Many arguments have been put forward in an attempt to find another explanation for the apparent link. Much more detailed studies were required to establish that smoking itself does indeed lead to an increased chance of developing lung cancer. However, although the data in Table 1 do not provide proof, they do support the belief that smoking cigarettes increases an individual's chance of contracting lung cancer.

1.4 Leukaemia clusters

There has been much publicity in the 1980s and early 1990s about the so-called clusters of childhood leukaemia cases around nuclear installations and, in particular, in the village of Seascale, near the Sellafield nuclear reprocessing plant in Cumbria, England. Look at Figures 3 and 4 which show extracts from two articles which were published in the *Independent* newspaper in January 1993. The first article, which appeared on 8 January, presents the case in favour of the view that the apparent cluster of childhood leukaemia cases is linked to the nuclear reprocessing plant. The second article, published three days later, looks at possible alternative explanations for the cluster.

Activity 6   It's obvious ... or is it?

Read the extracts now. What extra evidence is included in the second article in an attempt to produce an alternative view?

Sellafield cancer risk still above the average

By Liz Hunt and Tom Wilkie

YOUNG people living in the village of Seascale, near the Sellafield nuclear reprocessing plant in Cumbria, continue to run a greater risk of contracting leukaemia and other blood cancers, although scientists are still unable to explain why this is so.

A new report confirms the excess of leukaemias and non-Hodgkin's lymphomas in the village first identified by the Black report in 1984. It provides new evidence that these findings, for the years 1963 to 1983, were unlikely to be due to chance or the occurrence of a leukaemia "cluster".

In addition, the report shows that in the years following the Black report up to 1990, the incidence of the diseases in the 0-24 age group remained higher than both the national rates and those for the surrounding areas. This was "highly unlikely" to be due to chance, the report says.

When the Black report was published, some critics said that the excess of childhood cancers could be due to chance. Writing in tomorrow's issue of the *British Medical Journal*, the scientists say that "the accumulation of further data ... suggests this is not the correct explanation". They say the findings did not support or detract from the conclusions of the Gardner report in 1990 which found that cancers occurred in children whose fathers had high levels of radiation exposure before the child was conceived.

Figure 3   Newspaper article

New trails laid for childhood leukaemia hunt

YET MORE evidence has been produced suggesting that young people living in the village of Seascale, near the Sellafield nuclear reprocessing plant in Cumbria, run a greater risk of contracting leukaemia and other blood cancers.

A report by a team of epidemiologists headed by Gerald Draper of Oxford University and published in the *British Medical Journal*, confirms and updates the findings of the 1984 panel chaired by Sir Douglas Black, which concluded that there was indeed a "cluster" of childhood leukaemia cases and non-Hodgkin's lymphoma in the village. And it provides new evidence that the cluster, based on statistics for 1963 to 1983, was "highly unlikely" to have happened by chance.

The Draper analysis shows that in the six years after the Black report, the incidence of the diseases in the 0 to 24 age group remained higher than the national rate and those for the surrounding areas. But the study sheds no light on the cause. In particular, it neither supports nor disproves the controversial hypothesis, put forward in 1990 by Professor Martin Gardner of Southampton University, that there was a statistically significant relationship between a father's radiation dose in the six months before the conception of his child and the risk of the child subsequently developing leukaemia.

Professor Gardner's results have not been confirmed by other studies. For example, a survey of nuclear facilities in Ontario, Canada, published last year, found no increased risk of leukaemia in the children of fathers working at the plants. The Gardner results are further confounded by the fact that the cases were confined to Seascale, even though most of the workers at the Sellafield reprocessing plant live elsewhere.

These and other complications have led a number of scientists to suspect that other factors may be playing an equal, or even greater role. And although most epidemiologists believe that radiation and environmental pollution can be blamed for only a small fraction of childhood leukaemia, the emerging evidence suggests that Sellafield and other nuclear plants may be special cases, albeit with the potential to provide more general insights into the causes of the disease.

Sir Richard Doll, of Oxford University, says: "It does look as if there's some suggestion that, within the areas around many nuclear installations, there has been an excess of childhood leukaemia.

"But in this country and in Germany there has also been an excess reported in areas where nuclear stations were planned but never actually built. There is a lot to suggest that it is something characteristic of the area, rather than of the nuclear installation."

Two British scientists — Melvyn Greaves, director of the Leukaemia Research Fund Centre at the Institute for Cancer Research in London, and Leo Kinlen, director of the Cancer Epidemiology Research Group at Oxford — have set the leukaemia research world abuzz with a pair of related hypotheses, which trace childhood leukaemia to rare or unusual responses to infections early in life, including possible exposure to viruses.

Dr Greaves, in particular, has noted puzzling associations between a child's socio-economic status and the risk of falling victim to leukaemia. Unlike most other diseases, leukaemia poses a four times higher risk to children in developed countries than to those in the developing world.

This correlation was reinforced by a study released last year by the Office of Population Censuses and Surveys. A survey of 10,000 cases of childhood leukaemia and non-Hodgkin's lymphoma, led by Dr Draper's Childhood Cancer Research Group, concluded that the incidence in the wealthiest fifth of the British population was 10 to 15 per cent higher than in the poorest fifth.

Dr Greaves has suggested that the most frequently diagnosed type of childhood leukaemia — common acute lymphoblastic leukaemia — is caused by at least two spontaneous mutations in the child's white blood cells, which are important for the immune response. This model is consistent with the mechanisms now thought to cause other, better understood, types of cancer.

The first mutation occurs while the foetus is *in utero*, when the cells are immature and rapidly dividing, a situation that makes their

Figure 4 A second newspaper article

The articles, published in 1993, indicate that the issue of whether the Sellafield nuclear reprocessing plant is itself the cause of the childhood leukaemia cases in Seascale was still unresolved. The fact that the number of cases is small makes it difficult to establish whether the cluster could reasonably have arisen by chance. However, the existence of the cluster is an indication of a direction for investigation. This is another example of a pattern suggesting possible avenues for further study, but in this case no conclusion has yet been reached. (At least, no conclusion had been reached when this unit was written, in 1995.) Both articles refer to the cluster of childhood leukaemia cases in Seascale as being highly unlikely to have happened by chance, but neither claims that it could not be due to chance alone. Chance is the subject of the next section.

In this section, you have met several examples of data in health contexts where a possible relationship is suggested: the cluster of cholera cases near the Broad Street pump; the trend in the scatterplot of cigarette consumption plotted against lung cancer death rates; the cluster of childhood leukaemia cases in Seascale. All these patterns provide valuable information, not because they establish a causal link, *which they do not*, but because they suggest possible avenues for further investigation. In the cholera case, it was eventually found that the pump was to blame, and cigarette smoking is now generally accepted as a factor in lung cancer. However, no conclusion has yet been reached on the leukaemia cluster in Seascale.

Before moving on to the next section, take a few minutes to add responses to the sheet you started to complete in Activity 1. Think about *how* you are reading and responding to the materials for learning. For example, can you list the main points from this section?

Outcomes

After studying this section, you should be able to:
◇ appreciate how an insight into potential causes of illness and disease can be gained by looking at possible relationships in health data (Activities 2, 3, 5);
◇ draw a scatterplot of given data on graph paper and using your calculator (Activity 4);
◇ analyse critically arguments based on clusters (Activity 6);
◇ identify main points when reading a text (Activities 1 and 6).

2 *Chance*

Aims   The main aim of this section is to help you develop a better understanding of apparent patterns in data that can arise from chance alone.   ◇

In Section 1, you looked at several examples of patterns in health. At the level of an individual patient, nurse or doctor, general health patterns are not obvious. Only when case histories are collected together does it become possible to investigate *who* are becoming ill with *what* illnesses and *where* these are occurring. Section 1 focused particularly on the 'where' questions of health patterns, examining John Snow's cholera investigation and some of the clusters identified in occurrences of leukaemia. A crucial question underlying any analysis of clusters is whether such a cluster is a direct result of some factor or factors, or whether it is simply a chance occurrence. This is the central question of this section.

So you will be asked to consider just what sort of patterns tend to crop up naturally through chance alone. Research suggests that many people have a rather poor sense of this and typically underestimate the extent to which the outcomes of chance events tend to vary and cluster. As a result, they have a tendency to be more surprised and impressed by everyday 'coincidences' and associations than they should be. This subtle, but important, point was understood nearly two thousand years ago by the Greek essayist Plutarch when he wrote:

> It is no great wonder if, in the long process of time, while Fortune takes her course hither and thither, numerous coincidences should spontaneously occur.

If you are interested in exploring this issue further, there are two reader articles on the themes of belief, superstition and coincidence: one called 'Chance encounters' by John Allen Paulos, and another, an extract from *The Roots of Coincidence*, by Arthur Koestler.

The main purpose of this section, therefore, is to invite you to explore and develop your own understanding in this area of chance. You will be looking at statistical experiments and their corresponding *outcomes*. For example, an experiment might be a single toss of a coin, the outcome of which is either 'heads' or 'tails'. 'Heads' and 'tails' are assumed to be the only possible outcomes: the coin only comes down with its 'heads' side up, or its 'tails' side up—and not both. In general, an experiment may result in several possible outcomes, but only one of these outcomes can occur at a time.

Your calculator can generate some chance outcomes for you to explore any pattern in the numbers that result and to observe just how much variation and clustering appears to result *from chance alone*. After this section, you should be more aware of how patterns can arise by chance and be less ready to jump to (possibly unjustified) conclusions.

2.1 Random numbers

Imagine that you have ten identical balls, labelled 0 to 9, which are placed inside a bag and mixed thoroughly. Now, without looking inside the bag, pick a ball. Each ball should have an equal chance of being chosen. Write down the number of the ball chosen and replace it in the bag. Imagine you repeat the activity nineteen times. This would result in a run of twenty numbers from 0 to 9 each randomly chosen; such collections are called sets of random numbers.

Activity 7 An imaginary run of numbers

Write down a run of twenty numbers that you think might result from this exercise. *Please do not miss this activity out as you will shortly be asked to analyse the run of numbers that you have just thought up.*

Figure 5 shows a run of numbers that might be fairly typical of the run that you have written down.

$$4\ 1\ 9\ 0\ 6\ 3\ 2\ 7\ 8\ 3\ 4\ 6\ 5\ 9\ 3\ 2\ 0\ 8\ 5\ 7$$

Figure 5 Twenty 'made-up' random numbers

This might appear to be like a truly 'random' run of numbers. But how random is it really? Compare it with the run in Figure 6, which was generated using the randomizing function on the calculator.

$$9\ 0\ 8\ 3\ 1\ 8\ 8\ 6\ 1\ 1\ 1\ 4\ 6\ 6\ 8\ 7\ 8\ 2\ 9\ 2$$

Figure 6 Twenty 'calculator-generated' random numbers

At first sight, these two sequences of numbers may seem rather similar—they are both merely a run of digits from 0 to 9 with no special pattern. However, as you will see, the person who made up the twenty numbers in his head has actually imposed a greater degree of orderliness than was produced by the calculator.

SECTION 2 CHANCE

Activity 8 Successive pairs

Look at the two runs above and check whether any two consecutive numbers are the same. You should find that the person making up the 'random' sequence in his head has carefully avoided having two consecutive numbers the same, whereas the calculator has not.

Now check your own made-up run in this way. Have you tried to avoid repetitions?

Activity 9 Frequency count

Here is another check on orderliness. For each of the runs above, count the frequency of occurrence of each number. You should find that, for the made-up run, each number has occurred roughly twice but, for the calculator-generated run, one or two of the numbers occurred quite often, whereas others did not occur at all.

Now check your own made-up run in this way. Have each of your numbers come out with roughly the same frequency, or did the frequencies fluctuate as widely as for the calculator-generated run of numbers?

The comment on this activity suggested that, when people make up their own 'random' numbers, they tend to produce numbers with a much narrower range of frequencies than random numbers generated by a calculator (or computer or some mechanical process). This contrast (between 'made-up' and 'calculator-generated' random numbers) is even more evident when these figures are displayed in *frequency diagrams*, as shown in Figures 7 and 8. The possible values (0 to 9) are shown along the horizontal axis. The height of each bar corresponds to the frequency with which each value occurred—hence the name 'frequency diagram'.

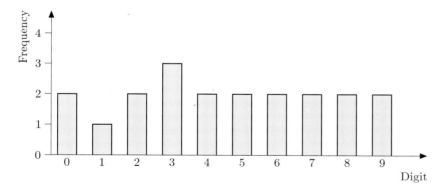

Figure 7 Frequency diagram showing the frequencies for the 'made-up' run

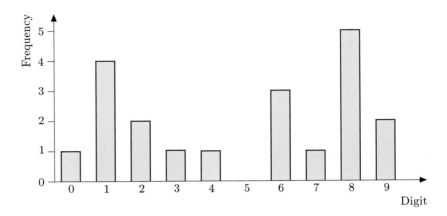

Figure 8 Frequency diagram showing the frequencies for the 'calculator-generated' run

The point of Activities 8 and 9 was to indicate that random events may produce outcomes which fluctuate more widely than most people expect. However, it is difficult to show this convincingly on the basis of a single run of twenty calculator-generated numbers. You really need to generate more than one run of random numbers on your calculator if you are to develop a good sense of the wide-ranging disorderliness of random events. You will be given help in how to do this shortly in the *Calculator Book*.

An important characteristic of a run of random numbers is that each of the possible numbers must be equally likely to occur. As you have already seen, because of natural variation, each sample will consist of numbers which may occur with unequal frequency, but do not let that obscure the underlying principle of randomness: namely, *each of the possible numbers has an equal chance of being selected every time.*

2.2 *Investigating variation in random numbers*

In this section, you will have the opportunity to use your calculator to generate some batches of random numbers for yourself. The process of using random numbers to see what might actually happen in practice is called *simulation*. Simulations involving random numbers will be used to investigate some patterns which may or may not occur by *chance*.

SECTION 2 CHANCE

Now work through Section 4.2 of Chapter 4 in the Calculator Book.

You have just recorded the results of generating sixty random digits using your calculator. Remember that, even though the outcomes were equally likely to occur, they will almost certainly not crop up equally often in practice. The object of the exercise was to get a sense of how much natural variation might be expected from repetitions of this sort of statistical experiment.

Activity 10 How much variation?

Look at the frequencies of the different digits 1, 2, 3, ..., 9, which *you* have generated, and write a few sentences describing the extent of the variation between the various outcomes. You may find it helpful to include a rough sketch showing a frequency diagram of your results, like the one shown in Figure 9, which used sixty random numbers generated by a course team member using the calculator.

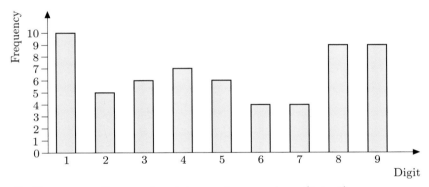

Figure 9 Frequency diagram for sixty random numbers (1 to 9)

The most frequent outcome occurred two and a half times as often as each of the least frequent outcomes ($10 = 2.5 \times 4$).

You might be wondering just how typical this degree of variation is. For example, if you calculate the ratio of maximum frequency to minimum frequency, based on your own data, do you get a result as large as 2.5?

Activity 11 Another simulation

Using the simulation that you have just carried out, record the following:

(a) maximum frequency;

(b) minimum frequency;

(c) the ratio $\dfrac{\text{maximum frequency}}{\text{minimum frequency}}$.

The calculator simulation was repeated a further nine times and the results of all ten runs are shown in Tables 2 and 3.

Table 2 Frequencies of the digits 1 to 9 in samples of size sixty

| | \multicolumn{9}{c}{Outcomes} |
	1	2	3	4	5	6	7	8	9
Run 1	10	5	6	7	6	4	4	9	9
Run 2	4	5	5	6	8	9	9	9	5
Run 3	7	9	9	8	6	3	5	5	8
Run 4	6	6	3	5	13	8	3	8	8
Run 5	10	8	8	6	5	4	6	7	6
Run 6	5	8	8	4	9	9	3	8	6
Run 7	7	7	9	11	8	3	5	5	5
Run 8	6	6	1	9	9	9	6	6	8
Run 9	8	5	7	6	13	4	5	7	5
Run 10	11	9	6	5	3	5	4	6	11

Table 3 Maximum and minimum frequencies

	Maximum frequency	Minimum frequency	Maximum frequency / Minimum frequency
Run 1	10	4	2.5
Run 2	9	4	2.25
Run 3	9	3	3
Run 4	13	3	4.33
Run 5	10	4	2.5
Run 6	9	3	3
Run 7	11	3	3.67
Run 8	9	1	9
Run 9	13	4	3.25
Run 10	11	3	3.67

These results seem to suggest that the first run (maximum frequency = 10, minimum frequency = 4) is by no means an untypical outcome and indeed, runs with even greater variation can occur.

The main points to emerge from this statistical investigation can be summarized as follows.

◇ Even when each outcome in a set of randomly occurring outcomes is equally likely, it is extremely unlikely that the different outcomes will occur with equal frequency.

◇ The degree of variation in frequency between the different possible outcomes is actually quite large—and probably larger than most people expect.

To end this subsection, spend a few minutes exploring how this investigation into variation in runs of random numbers relates to the health patterns that formed the central theme of the last section.

Imagine that in a particular year, there were a total of sixty recorded cases of leukaemia spread randomly over nine towns, all with roughly the same sized population. Further, suppose that one of those towns, which happened to be closest to a nuclear installation, was unlucky enough to account for thirteen of the sixty cases.

Activity 12   Cause or coincidence?

(a) How could you simulate the number of cases in each of the nine towns?

(b) How many cases of leukaemia, *on average*, might each town expect to record in that year?

(c) On the basis of the information in Table 3, do you think the citizens from the town which recorded thirteen cases have a particular cause for concern?

2.3 Hot spots

In the last subsection, you had the opportunity to use your calculator to generate simulated random data and investigate possible clusters or patterns in the results. However, although the calculator is quite a useful tool for handling small data sets, larger scale investigations require a computer. This subsection is centred around a video band which shows a computer animation of random occurrences. You will see how apparent 'clusters' can in fact occur purely by chance.

The central theme of this video band is 'clusters', in the context of a potentially fatal nineteenth-century disease, gallstones, and the parts of nineteenth-century Britain where incidences of gallstones tended to cluster. Some clusters occurred around large centres of population simply because there were more people in that part of the country. But, how can you know whether a cluster is a 'real' phenomenon or merely a chance occurrence. The video explores this question by means of a sequence of computer simulations. You are asked to compare the actual data with these simulations and look for similarities and differences between them. In order to do this, you may find it helpful to refer to Figure 10 which gives the actual data.

The book from which the data were taken is *Mortality in Mid-19th Century Britain* (1974) W. Farr and H. Ratcliffe, Gregg International, pp. 588–9.

Now watch 'Hot spots' on band 3 of DVD00107.

The video looked at some of the ways in which patterns in health can be investigated visually. In particular, it explored the incidence of deaths due to gallstones in Great Britain in the 1820s. The technique used was to generate a series of simulations based on an estimate of the population distribution at that time and on the basis of the overall known number of deaths in a typical year. These simulations were then compared visually with the actual data distribution about the regions.

UNIT 4 HEALTH

Activity 13 Making comparisons

In the video, you were asked the following questions.

◇ What features, if any, do the simulations have in common with the real data?

◇ What differences are there between the simulations and the real data?

You may have come to some general conclusions about this already. Now investigate these questions more systematically with the aid of the table below. Replay the DVD and view the simulated maps once more. As you look at each map, complete the appropriate column in Table 4. Work through each simulation in turn, comparing it with Figure 10.

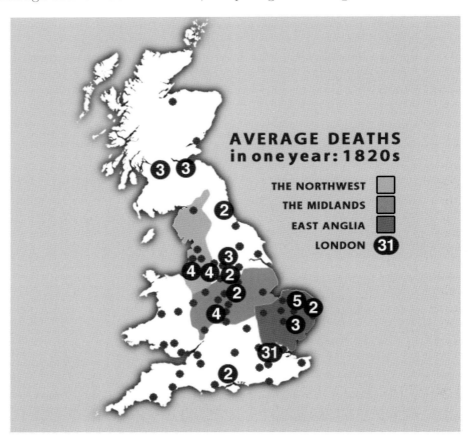

Figure 10

For each region, mark in the box how the simulation compares with the actual data. Where, for a particular region, the simulation seems to produce considerably more cases than the actual data, write M. If it produces considerably fewer, write F, and if there are roughly the same number of cases as the actual data, write S. Only do this very approximately—just indicate a general impression.

As you can see, the first one is done for you.

Table 4

	A	B	C	D	E	F	G	H
London	F							
Midlands and the North-west	S/M?							
East Anglia	F							
Scotland	M							
Wales	S/M?							
South and South-west (excluding London)	S							
North of England	S							

M = More   F = Fewer   S = Same

When you have completed the table, look across each completed row in turn. Are there consistent features for any of the regions? What does this suggest to you about the incidence of death due the gallstones in other regions? Do you think that the deaths occurred randomly overall in the population?

Two important points have emerged from the video. First, an apparent cluster may just be a chance occurrence. Second, apparent clusters tend to occur in centres of population, merely because more people live there. So, if you come across claims about clusters, do not automatically accept them at face value. There may be perfectly straightforward explanations for them.

The need to adjust figures to take this into account was discussed in the TV programme *Asthma and the Bean.*

A central message of this section has been about the usefulness of simulations when trying to interpret real data. They can give you an indication of whether or not patterns in the real data are likely to have arisen by chance. Were you aware of this message? How are you noting what you see as the main points?

Outcomes

After studying this section, you should be able to:

◇ generate random numbers using your calculator and to analyse the sort of patterns that crop up from chance alone (Activities 7–10);

◇ appreciate the degree of variation that can arise from chance alone and understand that this is the principal yardstick against which patterns in observed data should be judged (Activities 11, 12 and 13);

◇ feel more confident in using video material as part of your learning.

3 A pattern isn't proof

Aims The main aim of this short section is to allow you to step back and take an overview of the first two sections, focusing on random events and the significance of clusters. ◇

In Section 1, you looked at a variety of patterns in health. The first example looked at patterns of cholera deaths in Soho in the nineteenth century, as analysed by John Snow; later in the section, you looked at a more recent investigation into leukaemia clusters. The problem is that what appears to one person to be a significant cluster may just seem like a chance aggregation to another. Furthermore, even if there were no underlying cause, and the events under consideration were simply occurring randomly, a certain degree of clustering may well result *from chance alone*. So how can you distinguish 'real' clustering (due predominantly to a physical cause) from the sort that might occur due to chance variation? There is never any foolproof answer to this question. But there are a number of statistical tests, known collectively as *tests of significance*, which can help you to make these sorts of judgements. The decision is made on the basis of whether or not the magnitude of the cluster is 'beyond reasonable doubt'. The details of such tests of significance are outside the scope of MU120.

A central idea in this unit has been the exploration of random events and an investigation into the nature of random variation. In Subsection 2.2, you generated some random numbers on your calculator and saw what sort of 'patterns' tend to result from chance alone. In the video subsection (2.3), the process was repeated on a larger scale using a computer simulation. You may now have a better insight into just how much chance variation there can be and that you are sceptical of attributing automatic significance to apparently causal patterns in data.

But there is another issue worth considering, namely, what conclusions can sensibly be drawn when the observed patterns *are* more marked than anything expected to arise by chance? To take a specific example, assume that the clusters of childhood leukaemia cases close to nuclear installations are indeed unlikely to have arisen from chance alone. Does this therefore prove that the nuclear installations have caused the cases? The statistician's answer to this question is a resounding 'No!' As the title to this section states, 'a pattern isn't proof'. Even though the data may suggest that there is something going on here, the pattern alone does not prove that there is a cause and effect relationship. There may be any number of alternative explanations.

Activity 14 *There must be some simple explanation ...*

Taking the example described above, try to think up an alternative explanation for the clusters of childhood leukaemia cases close to nuclear installations other than that they were caused directly by radiation leaking from the nuclear plant itself. Think about how you could test this explanation.

It is always dangerous to jump to conclusions that are based on assumptions about cause and effect. If you think about things carefully, it is usually possible to come up with some plausible alternative explanations. The most likely alternative explanation here is that there is some other factor close to the installation which is causing the leukaemia cases. The medical explanation for this disease is not entirely clear. So this alternative factor could involve a variety of causes—contamination in the local water supply or in vegetable produce, for example.

Until there is a convincing medical explanation for the precise chain of events linking the suggested cause (some feature of the nuclear installation) with the observed consequence (childhood leukaemia cases in the locality), then causality has not been proved. In the absence of clear evidence for such a link, it is best to isolate and investigate each of the recognized alternatives and try to eliminate them. But even when all reasonable alternatives have been eliminated, there is always the possibility that the actual explanation lies with some other factor or combinations of factors that simply were not considered. For example, it is possible that some past event to which their parents or grandparents were exposed long before the nuclear plant was installed, led to the children having an increased risk of developing the disease.

The conclusion to be drawn from all of this seems to be that, even when patterns are evident, cause and effect relationships cannot be proved using statistics alone, and a medical explanation of the causal link is crucial to reaching firm conclusions. There are many classic textbook examples of patterns suggesting cause and effect relationships which are, in fact, ludicrous. For example, in a survey of villages in Germany, a link was discovered between the number of babies in each village and the corresponding number of storks nesting there. So are we to deduce that the storks are busily increasing the baby populations in these villages? Well, that is one possible explanation, but there is an alternative story, which seems a little more plausible. Villages with a larger number of babies tend to be larger villages with more houses. More houses means more chimneys, which, being favourite nesting places for storks, means more storks. So here is one possible explanation for the link: both the number of babies and the number of storks are linked to the size of the village.

Establishing the cause or causes of disease is a long, complex process of investigation, as is shown in the television programme *Asthma and the Bean*, which investigates the causes of exceptionally high incidence of asthma in one particular setting.

Outcomes

After studying this section, you should be able to:
◇ understand better some of the dangers of jumping to conclusions which are based on unwarranted assumptions about cause and effect (Activity 14).

4 Two tales of variation

Aims The main aim of this section is to discuss the strengths and weaknesses of various measures of spread. ◇

Interpretation of patterns is only meaningful if natural variation is taken into consideration—a 'pattern' can arise purely by chance, or it could be an indication of the presence of some underlying cause. Variation is an important concept in statistics, so this final section is devoted to reviewing the measures of spread that have been discussed so far (the range and the interquartile range discussed in *Unit 3*) and to introducing two other commonly used measures, the *standard deviation* and the *relative spread*.

4.1 A mother's tale

Shelley

My neighbour Maria recently had a baby girl, Shelley, who weighed 2490 grams at birth (about $5\frac{1}{2}$ pounds). After the initial congratulations were over, she confided to me her worry that Shelley was underweight. The other five babies born in the ward on the same day were all heavier at birth: their weights ranged from 2950 grams (about $6\frac{1}{2}$ pounds) to 3860 grams (about $8\frac{1}{2}$ pounds). She had heard that babies of low birth weight were at greater risk of suffering respiratory and other problems, so she was understandably concerned.

I tried to reassure her. I pointed out that the other five babies were all boys and that it was 'common knowledge' that boys were generally heavier at birth than girls. Lots of baby girls, I claimed, were as light or lighter than Shelley at birth. She refused to be reassured and so, to find out how much truth there was in my statements, I decided to collect some data. The questions I wanted to answer were: 'Are boys heavier than girls at birth?' and 'Was Shelley's birth weight unusually low?' And, I wondered to myself, 'Are the birth weights of girls and boys equally variable?'

The ward sister was very helpful. She gave me the birth weights to the nearest gram of the first 20 girls and the first 20 boys to be born in the ward in the previous month. The data are given in Table 5. (They have been sorted into ascending order.)

Table 5 Birth weights in grams of forty babies (twenty girls and twenty boys)

Girls	2082	2169	2438	2672	2808	2812	2862	2888	2983	3183
	3209	3277	3397	3490	3561	3565	3620	3673	3893	4188
Boys	2474	2559	2639	2653	2833	2971	3006	3218	3221	3294
	3438	3521	3591	3634	3705	3803	3942	4346	4479	4959

Activity 15 Was Shelley underweight?

(a) With the help of your calculator, sketch separate boxplots of the girls' and boys' birth weights given in Table 5. (You do not need to draw an accurate diagram: a rough sketch will do.)

(b) Do the data support the claim that boys are generally heavier than girls at birth?

(c) How does Shelley's birth weight compare with the weights of the 20 female babies?

(Do not clear the data from your calculator. You will need these figures again later in this section.)

More about Shelley

When I next visited Maria, I took along the data and my analysis of them. Although I was not able to stop her worrying altogether, she was somewhat reassured by the evidence.

What about the other question I was interested in? Are the birth weights of girls and boys equally variable?

Two measures of variation were introduced in *Unit 3*. The range of a batch of data is the numerical difference between the maximum and minimum values, *max* and *min*; and the interquartile range is the numerical difference between the quartiles, *Q3* and *Q1*. These two measures of spread are illustrated in Figure 11.

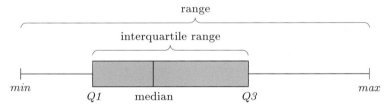

Figure 11

Activity 16 Measuring the spread

Calculate the range and the interquartile range of the girls' birth weights and of the boys' birth weights. Are the girls' or the boys' birth weights more widely spread?

Look again at the boxplots: you can see that the right whisker is much longer on the boxplot of the boys' birth weights than on the boxplot of the girls' birth weights. The birth weights of the heaviest boys are more widely spread than the birth weights of the heaviest girls. And on looking at the data (Table 5), you can see that there was one boy who weighed quite a lot more than the rest of the boys—4959 grams (nearly 11 pounds). The birth weight of this boy might account for the greater range in the birth weights of the boys. As you saw in *Unit 3*, the range is strongly influenced by extreme values. This was the reason for introducing an alternative measure of spread—the interquartile range—which is unaffected by unusually large or small values. However, since the highest 25% of values and the lowest 25% of values in a batch are ignored, a lot of information about a batch of data is lost when using the interquartile range. So now look at a third measure of spread.

The standard deviation

The range and the interquartile range, together with the median, provide a good deal of information about the distribution of the values in a batch of data; these can be represented visually in a boxplot. But suppose you decided to use the mean to represent a typical value in a batch, rather than the median: what measure of spread would it be appropriate to use to represent the variation in the batch?

The mean is an 'average' which takes every value in the batch into account—all are involved in its calculation. A commonly-used measure of spread which also uses every value in a batch is called the *standard deviation*. Whereas the interquartile range is usually used in conjunction with the median, the standard deviation is usually used with the mean. This is summarized in the table below.

Average	Measure of spread	Comment
Mean	Standard deviation	Uses all the values in its calculation
Median	Interquartile range	Does not use all the values in its calculation

The standard deviation is a measure of how widely the numbers in a batch are spread about the mean. As with the range and the interquartile range, the wider the spread of the batch values around the mean, then the greater the numerical value of the standard deviation.

The formula for the standard deviation is as follows.

$$\text{standard deviation} = \sqrt{\frac{\Sigma (x - \overline{x})^2}{n}}$$

This looks complicated, but the idea behind the formula is straightforward and, in practice, your calculator will do the actual calculations involved.

Equivalent formulas for the standard deviation are often used and you may see them elsewhere.

In words, the formula says that in order to work out the standard deviation for a batch of data, proceed as follows:
- ◇ calculate the mean of the batch, $\bar{x}$;
- ◇ for each number x in the batch, work out the numerical difference $x - \bar{x}$;
- ◇ square each difference, giving $(x - \bar{x})^2$;
- ◇ add together the entire list of squared differences, giving $\Sigma(x - \bar{x})^2$;
- ◇ divide your total by the number of values in the batch, n;
- ◇ finally, work out the square root of this last number;
- ◇ this gives the standard deviation.

$$\sqrt{\frac{\Sigma(x - \bar{x})^2}{n}}$$

Most calculators provide two different versions of the standard deviation—the one just described and a similar one in which, instead of dividing the sum of the squared deviations $\Sigma(x - \bar{x})^2$ by the batch size n, it is divided by $n - 1$. Both divisors are in common use. However, they are used in different circumstances. The discussion of when it is appropriate to use each divisor is beyond the scope of MU120. In this course, the divisor n will always be used: when you are asked for a standard deviation this is what is required. However, when reading about other people's results, you should always look to see which version of the formula has been used. In *Unit 5*, for instance, you will read about a situation where the person analysing some data has used the divisor $n - 1$, rather than n. The *Calculator Book* explains how to obtain the value of both versions of the standard deviation on your calculator—and which is which.

In some other courses, the divisor $n - 1$ is used.

Now work through Section 4.3 of Chapter 4 in the Calculator Book.

Activity 17 Finding the standard deviation

In Section 4.3 of Chapter 4 in the *Calculator Book*, you found how to calculate the standard deviation of a list of data. Use your calculator to find the standard deviation of the girls' and boys' birth weights separately, using Table 5 on page 29.

The standard deviation of the girls' birth weights is smaller than the standard deviation of the boys' birth weights (543.5 g compared with 655.0 g). So, for all three measures of spread that have been described—the range, the interquartile range and the standard deviation—the spread of the girls' birth weights is less than the spread of the boys' birth weights, but the difference is not very great. It is possible that girls' birth weights vary a little less than boys' birth weights, but these samples of birth weights do not provide strong evidence that this is the case. This small difference in spread might not be repeated in another sample or in the population at large. The difference might just be a feature of these particular data.

The effect of unusually high values on the range was discussed in *Unit 3*.

▶ What effect do unusually high values have on the standard deviation?

Activity 18 Mis-keyed data

Suppose that, when entering the data for the girls' birth weights in your calculator, you had entered the last weight as 41888 g instead of 4188 g, by mistake. What results would you have obtained for the range, the interquartile range and the standard deviation of the girls' birth weights? What does this tell you about the effect of unusually high values on the different measures of spread?

In this example, the interquartile range of the girls' birth weights is unaffected when one of the values is entered incorrectly as 41888 g instead of 4188 g. However, both the range and the standard deviation increase substantially. The range is defined to be the difference between the maximum and the minimum values, so this large increase is to be expected when an error of this type is made. The standard deviation changes from 543.5 g to 8471 g—a large increase when a single value is entered incorrectly.

But this effect is not just limited to a falsely large value. Large values of the actual data, which are a long way from the mean, increase the size of the standard deviation.

In *Unit 3*, you saw that the median is unaffected by unusually high values, whereas the mean is dragged upwards as the largest value increases. (Recall the effect on average pay of increasing the manager's pay at Troublefree Computers in *Unit 3*, Activity 8.) Similarly, the interquartile range is unaffected by an unusually large value, whereas the standard deviation increases as the largest value increases. These properties are summarized in the table below.

Mean and standard deviation	use all the values in their calculation and are ...	... strongly influenced by unusually large or small values in the data set.
Median and interquartile range	do not use extreme values in their calculation and are ...	... unaffected by unusually large or small values in the data set.

In summary, the story about birth weights of girls and boys was to help you to review the two measures of spread introduced in *Unit 3* (the range and the interquartile range) and to introduce and compare them with a third measure—the standard deviation.

4.2 A farmer's tale

Cows and calves

A farmer friend remarked to me recently about the birth of his son two weeks earlier than expected, 'Thank goodness cows are more predictable than women!' I was curious to know what he meant by this remark, so I asked him about it. He explained that cows are very predictable compared with women when it comes to giving birth. He could rely on his cows to give birth close to their expected date, so he could plan his work knowing there would not be any unexpected events.

I was intrigued by his assertion that 'cows are more predictable than women' and I decided to find out if there was any truth in it. To my surprise, I discovered that there was. I unearthed the information given in Table 6.

Table 6 Duration of pregnancies (weeks)

	Mean	Standard deviation
Women	39.2	2.6
Cows	40.2	0.7

Although the mean durations of human and bovine pregnancies are very similar, the standard deviation is much smaller for bovine pregnancies: the standard deviation of bovine pregnancies is only about a quarter of the corresponding figure for human pregnancies. At this point, I started to wonder whether the birth weights of calves are correspondingly less variable than the birth weights of babies. I speculated that a lot of the variation in the birth weights of babies is due to the variation in the lengths of pregnancies: early births usually result in smaller babies. So, since there is much less variation in the lengths of bovine pregnancies, surely there will be less variation in the birth weights of calves?

'Bovine' means 'relating to cattle'.

Table 7 contains the weights at birth of twenty Aberdeen Angus calves. The weights are given to an accuracy of one decimal place.

Data based on information supplied by the Meat and Livestock Commission.

Table 7 Birth weights of twenty calves (in kilograms)

48.0	23.7	25.5	45.5	38.0	32.8	47.2	33.6	32.6	31.2
21.9	38.7	33.2	32.6	34.8	40.2	35.8	30.5	27.8	39.9

Activity 19 *Comparing the spread*

(a) Use your calculator to find the mean, the median, the range, the interquartile range and the standard deviation of the birth weights of the calves.

(b) For the boys' birth weights, the comparable results from Activities 16 and 17 are:

$$\text{median} = 3366 \text{ g};$$
$$\text{range} = 2485 \text{ g};$$
$$\text{interquartile range} = 852 \text{ g};$$
$$\text{standard deviation} = 655 \text{ g}.$$

Compare the variation in the calves' birth weights with the variation in the boys' birth weights, taking care over the clarity of presentation of your results. Comment on your results.

Clearly, the variation in the calves' birth weights is much greater than the variation in the boys' birth weights. For example, the interquartile range of the calves' birth weights is 8.45 kg compared with an interquartile range of 852 g or 0.852 kg for the boys. However, you might expect the spread of the calves' birth weights to be greater: the median weight of the calves is 33.4 kg compared with a median weight of 3366 g or only 3.366 kg for the boys. So you would not expect to obtain an interquartile range as high as 8.45 kg for the birth weights of the boys! A variation of, say, 1 kg around an average of 3.4 kg for the boys would indicate much greater variability than a variation of 1 kg around an average of 34 kg.

For an average value which is ten times as large, you might expect the variation to be ten times as large if the variability is comparable. What is required is some means of assessing the *relative* variation of the birth weights of the calves and the boys.

One way of doing this is by dividing the interquartile range by the median (a measure of spread divided by a corresponding measure of size): the measure so obtained is called the *relative spread*. This measure is often given as a percentage.

$$\boxed{\text{relative spread} = \frac{\text{interquartile range}}{\text{median}} \times 100\%}$$

So, for instance, the relative spread of the boys' birth weights is

$$\frac{852}{3366} \times 100\% \simeq 25.3\%.$$

UNIT 4 HEALTH

Activity 20 *Comparing the relative spreads: calves and boys*

Find the relative spread of the birth weights of the calves. Is the variation in the birth weights of the calves relatively less than the variation in the boys' birth weights?

So the data do not support the hypothesis that there would be less variation in the birth weights of calves than in the birth weights of babies. But, presumably, the sample of calves' birth weights included the birth weights of both male and female calves. You compared the relative spread of the calves' birth weights with the relative spread of the boys' birth weights. Would you have obtained the same result if you had compared the relative spread of the calves' birth weights with the relative spread of the girls' birth weights, or with the relative spread of the forty babies' birth weights?

Activity 21 *Comparing the relative spreads: calves and girls*

(a) Find the relative spread of the birth weights of the girls. Is the variation in the birth weights of the girls relatively less than or greater than the variation in the calves' birth weights?

(b) Combine the birth weights of the girls and boys into a single batch of forty babies' birth weights. Find the relative spread of the birth weights of the babies. Is the variation in the birth weights of the babies relatively less than or greater than the variation in the birth weights of the calves?

In this section, various measures of spread have been discussed. The range is strongly influenced by extreme values in a batch of data, so it can fail to give an accurate indication of the spread of the bulk of the numbers in a batch of data. On the other hand, the interquartile range does not depend at all on the extreme values, but ignores the top 25% and the bottom 25% of numbers in a batch. These two measures are commonly employed when the average used is the median.

The standard deviation is a measure of spread that uses all the numbers in a batch; it is usually used in conjunction with the mean—the average which also uses all the numbers. However, like the range, the standard deviation is strongly influenced by unusually large or small numbers.

When the sizes of typical values in two or more batches are very different—for instance, when the numbers in one batch are several times as large as the numbers in another—you need to compare the relative variation of the numbers in the batches. A measure of spread which takes the size of the numbers into account is the relative spread, which depends on the interquartile range and the median.

Outcomes

After studying this section, you should be able to:

◇ use your calculator to find the standard deviation of a batch of data (Activities 17 to 19);

◇ find the relative spread of a batch of data (Activities 20 and 21);

◇ calculate and describe some of the properties of different measures of spread in their use for comparing data sets (Activities 15 to 19);

◇ read academic text for a variety of purposes more proficiently.

Unit summary and outcomes

In this unit, you have looked at the representations and interpretation of patterns in data, particularly as they may occur in the context of health and illness. Several of these contexts were historical. One example involved the map drawn by John Snow in his investigation of the cholera epidemic in London. Another, also from nineteenth-century Britain, was Florence Nightingale's development of innovative graphical representations designed to create a climate of opinion that enabled vital hospital reforms to be carried out.

The unit explored two dilemmas which tend to underpin all decisions that are based on the interpretation of patterns in data. First, are the patterns sufficiently distinct for us to be able to conclude that they may be due to factors other than chance variation? Second, in situations where a pattern is clearly evident, what might the causal factor or factors be?

Thinking about your progress

You now have quite a lot of experience at using your Learning File to help you learn effectively. One of the things you have been encouraged to do is identify and reword key points from a text. Now would be a good time to reflect on what you find helpful in doing this—for example, referring to the 'aims' statement or using the outcomes list. Summarize your ideas on the Learning File response sheet you started at the beginning of the unit. Are you intending to do something different for *Unit 5*?

Activity 22 Looking back

As in previous units, record your notes on key terms introduced, on the *Unit 4* Handbook activity sheet. So, as review work, go back through the unit and record your notes on terms such as random sample, scatterplot, chance occurrence, simulation, frequency diagram, standard deviation, relative spread. (Your list may not be identical to this.)

Also, take the time to reflect on the progress you have made in learning mathematics. What topics in this unit have you found straightforward? What have you found difficult? Write down what you feel you have gained from studying this unit—for example, a skill that you have improved or an understanding gained of some idea or technique.

Write down one example of something that caused you difficulty and on which you feel you need to spend more time.

◇ If you have identified some aspect of the work in this unit that is causing you real concern, how are you going to overcome it?

◇ If you identified something in one of the earlier units that required action on your part, what did you try then?

UNIT SUMMARY AND OUTCOMES

- ◇ Did you remember to record the results of your action in your Learning File?
- ◇ Use your experience from that occasion to help you to decide what to do now.

Finally, look back to your various responses for Activity 1. How well are you doing at 'reading for learning'? Summarize your thoughts before moving on, and check your learning against the outcomes below.

Outcomes

You should now be able to:

- ◇ appreciate how an insight into potential causes of illness and disease can be gained by looking at patterns in health data;
- ◇ discuss why patterns can arise purely by chance and be aware that the existence of a pattern does not prove the existence of a cause and effect relationship;
- ◇ generate random numbers using your calculator;
- ◇ use your calculator to find the standard deviation of a batch of data;
- ◇ find the relative spread of a batch of data;
- ◇ describe the properties (including advantages and disadvantages) of different measures of spread.

Comments on Activities

Activity 1

There are no comments on this activity.

Activity 2

There were far more cases in the neighbourhood of the Broad Street pump than near the other pumps. This fact alone suggests the possibility of the Broad Street pump being implicated in the outbreak.

Activity 3

See the comments after the activity.

Activity 4

When you work through Section 4.1 of Chapter 4 in the *Calculator Book*, check your scatterplot against the display on the screen of your calculator. (The complete scatterplot is shown in Figure 2 of the main text.)

Activity 5

See the comments after the activity.

Activity 6

First, the second article observes that the cases of leukaemia in the children of fathers working at Seascale were confined to Seascale, even though most of the workers lived elsewhere. Second, no similar increased risk of leukaemia was found in a survey of nuclear facilities in Ontario. Third, it observes that there is evidence of an increased risk in areas where nuclear stations have been planned but never built: so the cause might be something characteristic of an area suitable for a nuclear installation rather than the installation itself.

Activity 7

There are no comments on this activity.

Activity 8

Analysing the occurrence of pairs produces the following results. No two consecutive numbers were equal in the made-up run.

As shown below, in the calculator-generated run, two consecutive numbers were equal on four occasions. (The two consecutive pairs of 1s actually produce a triple.)

9, 0, 8, 3, 1, 8, 8, 6, 1, 1, 1,
                   pair   pair pair

4, 6, 6, 8, 7, 8, 2, 9, 2
   pair

Activity 9

A count of the numbers for each run produces the following results.

Number	Frequency for made-up run	Frequency for calculator run
0	2	1
1	1	4
2	2	2
3	3	1
4	2	1
5	2	0
6	2	3
7	2	1
8	2	5
9	2	2

As can be seen from the table, the frequencies range from zero to five for the numbers generated by the calculator, but the made-up numbers have a narrower range from one to three. In other words, in respect of their frequency of occurrence, the made-up numbers are much more orderly than the calculator-generated numbers.

Activity 10

There was considerable variation in the frequencies of the sixty numbers which were generated (by a course team member using the calculator). The outcome with the largest frequency was 1, which occurred ten times. Outcomes 6 and 7 occurred only four times each. Your results will obviously differ.

Activity 11

See the comments after the activity.

Activity 12

(a) Each town could be identified with a different number from 1 to 9. Sixty random whole numbers in the range 1 to 9 inclusive could then be generated and a count made of how many times each number crops up (exactly as was done in Activity 10).

(b) If the sixty cases were spread evenly across the nine towns, you would expect 60 ÷ 9 or roughly six or seven cases per town that year.

(c) The actual number of cases in the town which recorded thirteen cases is roughly twice as many as you might expect to get by chance alone. However, as you can see from Table 3, out of the ten simulated runs, two of them produced maximum frequencies as large as thirteen. The citizens of the town in question might conclude, therefore, that this result is within the range of what they might expect from chance alone. However, they would be advised to monitor the results carefully over subsequent years just in case a similar pattern recurs, in which case there would be a much greater cause for concern.

Activity 13

Our observations can be summarized as follows.

	A	B	C	D
London	F	F	F	F
Midlands and the North-west	S/M?	M?	S/M?	M
East Anglia	F	F	F	F
Scotland	M	M	M	M
Wales	S/M?	S/M?	S/M?	S
South and South-west (excluding London)	S	S	S	S
North of England	S	S	M	S

	E	F	G	H
London	F	F	F	F
Midlands and the North-west	S/M?	S?	M?	M?
East Anglia	F	F	F	F
Scotland	S	M	M	M
Wales	S/M?	S	S?	S
South and South-west (excluding London)	S	M?	S	S
North of England	M?	S	M?	S

The table suggests the following general features.

In London and East Anglia, the simulation consistently produced fewer cases than were actually recorded, whereas in Scotland, in all but one case, the reverse was true. This seems to suggest that people in London and East Anglia were more likely, and people in Scotland were less likely, to die from gallstones than in other regions.

The large number of actual cases recorded in the Midlands and the North-west may have seemed to support believing there to be a greater risk in these regions. However, the simulations suggest that this is not the case. The reason for the large number of cases in these areas may simply be that more people lived there than in other parts of Britain.

Activity 14

Sites suitable for nuclear installation might have an environmental factor which causes leukaemia (for example, near the sea), or something else to do with the plant's construction or operation may be responsible.

Activity 15

(a) Boxplots of the girls' birth weights and the boys' birth weights are shown below.

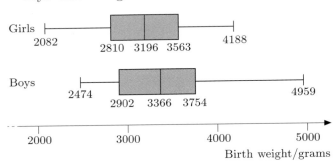

Boxplots of girls' and boys' birth weights (in grams)

(b) Each of the five summary statistics marked on the boxplots is higher for the boys than the corresponding statistic for the girls. The lightest girl weighed less than the lightest boy at birth, the lower quartile of the girls' birth weights was less than the lower quartile of the boys' birth weights, and so on. The data do provide some evidence to support the statement that boys are generally heavier than girls at birth. However, there does not appear to be a very large difference.

(c) Of the twenty girls, three were lighter than Shelley at birth, so it appears that, although Shelley's birth weight was below average, it was not so low as to be remarkable. If this sample of twenty girls' birth weights is typical, then quite a few girls will weigh less than Shelley did at birth.

Activity 16

For the girls, the range and interquartile range are as follows:

$$\text{range} = max - min = 4188\,\text{g} - 2082\,\text{g}$$
$$= 2106\,\text{g};$$
$$\text{interquartile range} = Q3 - Q1$$
$$= 3563\,\text{g} - 2810\,\text{g}$$
$$= 753\,\text{g}.$$

For the boys,

$$\text{range} = 4959\,\text{g} - 2474\,\text{g} = 2485\,\text{g};$$
$$\text{interquartile range} = 3754\,\text{g} - 2902\,\text{g}$$
$$= 852\,\text{g}.$$

So there is very little difference between the spread of the girls' birth weights and the spread of the boys' birth weights—at least, when the measure of spread used is either the range or the interquartile range. The spread is slightly greater for the boys' birth weights whichever measure is used.

Activity 17

The standard deviations of the girls' and boys' birth weights are, respectively, 543.5 g and 655.0 g. This shows that the boys' weights are more widely spread than the girls'.

Activity 18

The values of the three measures of spread are as follows:

$$\text{range} = 41888\,\text{g} - 2082\,\text{g} = 39806\,\text{g};$$
$$\text{interquartile range} = 3563\,\text{g} - 2810\,\text{g} = 753\,\text{g};$$
$$\text{standard deviation} = 8471.324067\,\text{g} \simeq 8471\,\text{g}.$$

The effect is discussed after the activity.

Activity 19

(a) For the birth weights of the calves,

$$\text{mean} = 34.7 \, \text{kg};$$
$$\text{median} = 33.4 \, \text{kg};$$
$$\text{range} = max - min = 48.0 \, \text{kg} - 21.9 \, \text{kg}$$
$$= 26.1 \, \text{kg};$$
$$\text{interquartile range} = Q3 - Q1$$
$$= 39.3 \, \text{kg} - 30.85 \, \text{kg}$$
$$= 8.45 \, \text{kg};$$
$$\text{standard deviation} \simeq 7.1 \, \text{kg}.$$

(b) See the comments after the activity.

Activity 20

The relative spread of the birth weights of the calves is

$$\frac{8.45}{33.4} \times 100\% \simeq 25\%.$$

So the relative spread of the calves' birth weights is almost exactly the same as the relative spread of the birth weights of the boys. This means there is relatively the same amount of variation in the calves' birth weights as in the birth weights of the boys.

Activity 21

(a) For the birth weights of the girls,

$$\text{interquartile range} = 753 \, \text{g};$$
$$\text{median} = 3196 \, \text{g}.$$

So the relative spread of the birth weights of the girls is

$$\frac{753}{3196} \times 100\% \simeq 23.6\%.$$

Therefore, the relative spread of the girls' birth weights is very slightly less than the relative spread of the birth weights of the calves.

(b) For the combined sample of forty babies' birth weights:

$$\text{interquartile range} = Q3 - Q1$$
$$= 3627 \, \text{g} - 2822.5 \, \text{g}$$
$$= 804.5 \, \text{g};$$
$$\text{median} = 3249 \, \text{g}.$$

So the relative spread is

$$\frac{804.5}{3249} \times 100\% \simeq 24.8\%.$$

Therefore, the relative spread of the babies' birth weights is very similar to the relative spread of the calves' birth weights. There is the same amount of spread relatively in the birth weights of the babies as in the birth weights of the calves.

Activity 22

There are no comments on this activity.

Acknowledgements

Grateful acknowledgement is made to the following sources for permission to reproduce material in this unit:

Text

Hunt, L. and Wilkie, T. 'Sellafield cancer risk still above average,' *Independent*, 8.1.1993; Balter, M. 'New trails laid for childhood leukaemia hunt', *Independent*, 11.1.1993.

Photograph

p.8: John Snow: photo: by permission of the British Library (7560e67)

Cover

Guillemots: RSPB Photo Library; Sellafield newspaper headline: *Independent*, 8.1.1993; other photographs: Mike Levers, Photographic Department, The Open University.

Index

cause   23
cause and effect relationships   7, 26
chance occurrence   38
clusters   13
coincidence   23

frequency count   19
frequency diagrams   19, 38

interquartile range   30

max   30
min   30

paired data   11
patterns   8, 12, 13

Q1   30
Q3   30

random events   26
random numbers   18
random sample   38
random variation   26
relative spread   35, 38

scatter diagram   11
scatterplot   11, 38
simulation   20, 38
spread   30
standard deviation   31, 38
statistical experiments   17

variation   20, 29